CONTENTS

PART 1. GENERAL

PART 2. PRODUCTS

PART 3. EXECUTION

(This publication is adapted from the *Construction Criteria Base* of the United States government which is in the public domain, is authorized for unlimited distribution, and is not copyrighted.)

PART 1 GENERAL

1.1 REFERENCES. The publications listed below form a part of this specification to the extent referenced. The publications are referred to within the text by the basic designation only.

<u>AMERICAN CONCRETE INSTITUTE INTERNATIONAL (ACI)</u>

ACI 318 (2011; Errata 1 2011; Errata 2 2012; Errata 3-4 2013) Building Code Requirements for Structural Concrete and Commentary

ACI 318M (2011; Errata 2013) Building Code Requirements for Structural Concrete & Commentary

ACI 530/530.1 (2011; Errata 2011; Errata 2013) Building Code Requirements and Specification for Masonry Structures and Related Commentaries

ACI SP-66 (2004) ACI Detailing Manual

<u>ASTM INTERNATIONAL (ASTM)</u>

ASTM A153/A153M (2009) Standard Specification for Zinc Coating (Hot-Dip) on Iron and Steel Hardware

ASTM A167 (1999; R 2009) Standard Specification for Stainless and Heat-Resisting Chromium-Nickel Steel Plate, Sheet, and Strip

ASTM A615/A615M (2012) Standard Specification for Deformed and Plain Carbon-Steel Bars for Concrete Reinforcement

ASTM A641/A641M (2009a) Standard Specification for Zinc-Coated (Galvanized) Carbon Steel Wire

ASTM A82/A82M (2007) Standard Specification for Steel Wire, Plain, for Concrete Reinforcement

ASTM B370 (2012) Standard Specification for Copper Sheet and Strip for Building Construction

ASTM B633 (2013) Standard Specification for Electrodeposited Coatings of Zinc on Iron and Steel

ASTM C1019 (2013) Standard Test Method for Sampling and Testing Grout

ASTM C1072 (2013) Standard Test Method for Measurement of Masonry Flexural Bond Strength

ASTM C1142 (1995; R 2013) Standard Specification for Extended Life Mortar for Unit Masonry

ASTM C129 (2011) Standard Specification for Nonloadbearing Concrete Masonry Units

ASTM C140/C140M (2013a) Standard Test Methods for Sampling and Testing Concrete Masonry Units and Related Units

ASTM C144 (2011) Standard Specification for Aggregate for Masonry Mortar

ASTM C150/C150M (2012) Standard Specification for Portland Cement

ASTM C207 (2006; R 2011) Standard Specification for Hydrated Lime for Masonry Purposes

ASTM C216 (2013) Facing Brick (Solid Masonry Units Made from Clay or Shale)

ASTM C27 (1998; R 2008) Fireclay and High-Alumina Refractory Brick

ASTM C270 (2012a) Standard Specification for Mortar for Unit Masonry

ASTM C315 (2007; R 2011) Clay Flue Linings

ASTM C476 (2010) Standard Specification for Grout for Masonry

ASTM C494/C494M (2013) Standard Specification for Chemical Admixtures for Concrete

ASTM C55 (2011) Concrete Brick

ASTM C593 (2006; R 2011) Fly Ash and Other Pozzolans for Use with Lime for Soil Stabilization

ASTM C62 (2013) Building Brick (Solid Masonry Units Made from Clay or Shale)

ASTM C641 (2009) Staining Materials in Lightweight Concrete Aggregates

ASTM C652 (2013) Hollow Brick (Hollow Masonry Units Made from Clay or Shale)

ASTM C67 (2013) Standard Test Methods for Sampling and Testing Brick and Structural Clay Tile

ASTM C73 (2010) Calcium Silicate Brick (Sand-Lime Brick)

ASTM C780 (2012a) Preconstruction and Construction Evaluation of Mortars for Plain and Reinforced Unit Masonry

ASTM C90 (2013) Loadbearing Concrete Masonry Units

ASTM C91/C91M (2012) Standard Specification for Masonry Cement

ASTM C94/C94M (2013a) Standard Specification for Ready-Mixed Concrete

ASTM C989/C989M (2012a) Standard Specification for Slag Cement for Use in Concrete and Mortars

ASTM D1972 (1997; R 2005) Standard Practice for Generic Marking of Plastic Products

ASTM D2000 (2012) Standard Classification System for Rubber Products in Automotive Applications

ASTM D2240 (2005; R 2010) Standard Test Method for Rubber Property - Durometer Hardness

ASTM D2287 (2012) Nonrigid Vinyl Chloride Polymer and Copolymer Molding and Extrusion Compounds

ASTM E119 (2012a) Standard Test Methods for Fire Tests of Building Construction and Materials

ASTM E2129 (2010) Standard Practice for Data Collection for Sustainability Assessment of Building Products

ASTM E514/E514M (2011) Standard Test Method for Water Penetration and Leakage Through Masonry

INTERNATIONAL CODE COUNCIL (ICC)

ICC IBC (2012) International Building Code

U.S. DEPARTMENT OF DEFENSE (DOD)

UFC 3-310-04 (2012) Seismic Design for Buildings

U.S. GREEN BUILDING COUNCIL (USGBC)

LEED NC (2009) Leadership in Energy and Environmental Design(tm) New Construction Rating System

1.2 SYSTEM DESCRIPTION

1.2.1 LOCAL/REGIONAL MATERIALS. [Use materials or products extracted, harvested, or recovered, as well as manufactured, within a [800] [_____] km [500] [_____] mile radius from the project site, if available from a minimum of three sources.] [See Section 01 33 29 LEED(tm) DOCUMENTATION for cumulative total local material requirements. Masonry materials may be locally available.] Submit documentation indicating distance between manufacturing facility and the project site, and distance of raw material origin from the project site. Indicate relative dollar value of local/regional materials to total dollar value of products included in the project.

1.2.2 ENVIRONMENTAL DATA. Submit manufacturer's descriptive data. Documentation indicating percentage of post-industrial and post-consumer recycled content per unit of product. Indicate relative dollar value of recycled content products to total dollar value of products included in project. [Submit Table 1 of ASTM E2129 for the following products: [_____].]

1.2.3 PLASTIC IDENTIFICATION. Verify that plastic products to be incorporated into the project are labeled in accordance with ASTM D1972. Where products are not labeled, submit product data indicating polymeric information in the Operation and Maintenance Manual.

Type
1 Polyethylene Terephthalate (PET, PETE)
2 High Density Polyethylene (HDPE)
3 Vinyl (Polyvinyl Chloride or PVC)
4 Low Density Polyethylene (LDPE)
5 Polypropylene (PP)
6 Polystyrene (PS)
7. Other. Use of this code indicates that the package in question is made with a resin other than the six listed above, or is made of more than one resin listed above, and used in a multi-layer combination.

1.2.4 DESIGN REQUIREMENTS

1.2.4.1 UNIT STRENGTH METHOD. Compute compressive strength of masonry system "Unit Strength Method", ACI 530/530.1. Submit calculations and certifications of unit and mortar strength.

1.2.4.2 SEISMIC REQUIREMENT. In addition to design requirements of ICC IBC, provide additional seismic reinforcement [in accordance with UFC 3-310-04] [as detailed on [the drawings] [sketches [_____] which are attached at the rear of this section]]. The total minimum reinforcing percentage for structural walls shall be 0.20 percent and non-structural walls shall be 0.15 percent. The maximum spacing of reinforcing bars shall be as follows:

Wall Type	Vertical	Horizontal
Structural	0.609 m 24 inches	1.219 m 48 inches
Non-structural	1.219 m 48 inches	2.032 m 80 inches

Bond beams are required at the top of footings, at the bottom and top of openings at roof and floor levels, and at the top of parapet walls.

1.2.4.3 SPECIAL INSPECTION. Perform special inspections and testing for seismic-resisting systems and components in accordance with UFC 3-310-04 SEISMIC DESIGN FOR BUILDINGS and Section SPECIAL INSPECTIONS.

1.2.4.4 MASONRY STRENGTH. Determine masonry strength in accordance with ACI 530/530.1; submit test reports on three prisms as specified in ACI 530/530.1. The cost of testing shall be paid by the Contractor.

1.2.5 ADDITIONAL REQUIREMENTS

a. Maintain at least one spare vibrator on site at all times.
b. Provide bracing and scaffolding necessary for masonry work. Design bracing to resist wind pressure as required by local code.

1.2.6 METRICATION. The Contractor has the option to use either hard metric or substitute inch-pound (soft-metric) CMU products. If the Contractor decides to substitute inch-pound CMU products, meet the following additional requirements:

a. The metric dimensions indicated on the drawings shall not be altered to accommodate inch-pound CMU products either horizontally or vertically. The 100 mm building module shall be maintained, except for the CMU products themselves.
b. Mortar joint widths shall be maintained as specified.

c. Rebars shall not be cut, bent or eliminated to fit into the inch-pound CMU products module.

d. Brick and inch-pound CMU products shall not be reduced in size by more than one-third (1/3) in height and one-half (1/2) in length. Cut CMU products shall not be located at ends of walls, corners, and other openings.

e. Cut, exposed brick and CMU products shall be held to a minimum and located where they would have the least impact on the architectural aesthetic goals of the facility.

f. Other building components, built into the CMU products, such as window frames, door frames, louvers, grilles, fire dampers, etc., that are required to be metric, shall remain metric.

1.3 SUBMITTALS. Owner approval is required for submittals with a "G" designation; submittals not having a "G" designation are for [Contractor Quality Control approval.] [information only. When used, a designation following the "G" designation identifies the office that will review the submittal for the Owner.] Submit the following in accordance with Section SUBMITTAL PROCEDURES:

SD-02 Shop Drawings
Detail Drawings[; G][; G, [_____]]

SD-03 Product Data
[Local/Regional Materials; (LEED NC)]
[Environmental Data]
Clay or Shale Brick[; G][; G, [_____]]; (LEED NC)
Cement[; G][; G, [_____]]; (LEED NC)
Insulation[; G][; G, [_____]]
Cold Weather Installation[; G][; G, [_____]]
Salvaged Brick[; G][; G, [_____]]; (LEED NC)
Water-Repellant Admixture[; G][; G, [_____]]

SD-04 Samples

Concrete Masonry Units (CMU)[; G][; G, [_____]]

Concrete Brick[; G][; G, [_____]]

Stone Items[; G][; G, [_____]]

Clay or Shale Brick[; G][; G, [_____]]

Anchors, Ties, and Bar Positioners[; G][; G, [_____]]

Expansion-Joint Materials[; G][; G, [_____]]

Joint Reinforcement[; G][; G, [_____]]

Insulation[; G][; G, [_____]]

Portable Panel[; G][; G, [_____]]

SD-05 Design Data

Pre-mixed Mortar[; G][; G, [_____]]

Unit Strength Method[; G][; G, [_____]]

SD-06 Test Reports

Efflorescence Test[; G][; G, [_____]]

Field Testing of Mortar[; G][; G, [_____]]

Field Testing of Grout[; G][; G, [_____]]

Prism tests[; G][; G, [_____]]

Masonry Cement[; G][; G, [_____]]

Fire-rated CMU[; G][; G, [_____]]

Masonry Inspector Qualifications[; G][; G, [_____]]

Single-Wythe Masonry Wall Water Penetration Test

SD-07 Certificates

Clay or Shale Brick

Concrete Brick

Concrete Masonry Units (CMU)

Anchors, Ties, and Bar Positioners

Expansion-Joint Materials

Joint Reinforcement

Masonry Cement

Insulation

Precast Concrete Items

Admixtures for Masonry Mortar

Admixtures for Grout

Insulation

Contamination

SD-08 Manufacturer's Instructions

Masonry Cement

SD-10 Operation and Maintenance Data

Take-Back Program

1.4 QUALITY ASSURANCE

1.4.1 APPEARANCE. [Manufacture bricks at one time and from the same batch.] Blend all brick to produce a uniform appearance when installed. An observable "banding" or "layering" of colors or textures caused by improperly mixed brick is unacceptable.

1.4.2 CONTAMINATION. When using bricks containing contaminated soil, supplier shall certify that the hazardous waste is neutralized by the manufacturing process and that no additional pollutants will be released, or that the product is free from hazardous contaminants.

1.4.3 SAMPLE MASONRY PANELS. After material samples are approved and prior to starting masonry work, construct a portable panel of clay or shale brick and sample masonry panels for each type and color of masonry required. At least 48 hours prior to constructing the sample panel or panels, submit written notification to the Owner. Submit one panel of clay or shale brick, 600 by 600 mm 2 by 2 feet, containing approximately 24 brick facings to establish range of color and texture. Sample panels shall not be built in, or as part of the structure, but shall be located where directed.

1.4.3.1 CONFIGURATION. Panels shall be L-shaped or otherwise configured to represent all of the wall elements. Panels shall be of the size necessary to demonstrate the acceptable level of workmanship for each type of masonry represented on the project. The minimum size of a straight panel or a leg of an L-shaped panel shall be 2.5 m 8 feet long by [1.2] [1.8] m [4] [6] feet high.

1.4.3.2 COMPOSITION. Panels shall show full color range, texture, and bond pattern of the masonry work. The Contractor's method for mortar joint tooling; grouting of reinforced vertical cores, collar joints, bond beams, and lintels; positioning, securing, and lapping of reinforcing steel; positioning and lapping of joint reinforcement (including prefabricated corners); and cleaning of masonry work shall be demonstrated during the construction of the panels. Installation or application procedures for anchors, wall ties, CMU control joints, brick expansion joints, insulation, flashing, brick soldier, row lock courses and weep holes shall be shown in the sample panels. The panels shall contain [a masonry bonded corner] [a stacked bond corner] that includes a bond beam corner. Panels shall show [parging] [and] [installation of electrical boxes and conduit]. Panels that represent reinforced masonry shall contain a 600 by 600 mm 2 by 2 foot opening placed at least 600 mm 2 feet above the panel base and 600 mm 2 feet away from all free edges, corners, and control joints. Required reinforcing shall be provided around this opening as well as at wall corners and control joints.

1.4.3.3 CONSTRUCTION METHOD. Where anchored veneer walls are required, demonstrate and receive approval for the method of construction; i.e., either bring up the two wythes together or separately, with the insulation and appropriate ties placed within the specified tolerances across the cavity. Temporary provisions shall be demonstrated to preclude mortar or grout droppings in the cavity and to provide a clear open air space of the dimensions shown on the drawings. Where masonry is to be grouted, demonstrate and receive approval on the method that will be used to bring up the masonry wythes; support the reinforcing bars; and grout cells, bond beams, lintels, and collar joints using the requirements specified herein. If sealer is specified to be applied to the masonry units, sealer shall be applied to the sample panels. Panels shall be built on a properly designed concrete foundation.

1.4.3.4 USAGE. The completed panels shall be used as the standard of workmanship for the type of masonry represented. Masonry work shall not commence until the sample panel for that type of masonry construction has been completed and approved. Panels shall be protected from the weather and construction operations until the masonry work has been completed and approved. After completion of the work, the sample panels, including all foundation concrete, shall become the property of the Contractor and shall be removed from the construction site.

1.4.4 MASONRY INSPECTOR QUALIFICATIONS. A qualified masonry inspector approved by the Owner shall perform inspection of the masonry work. Minimum qualifications for the masonry inspector shall be 5 years of reinforced masonry inspection experience or acceptance by a State, municipality, or other governmental body having a program of examining and certifying inspectors for reinforced masonry construction. The masonry inspector shall be present during preparation of masonry prisms, sampling and placing of masonry units, placement of reinforcement (including placement of dowels in footings and foundation walls), inspection of grout space, immediately prior to closing of cleanouts, and during grouting operations. The masonry inspector shall assure compliance with the drawings and specifications. The masonry inspector shall keep a complete record of all inspections and shall submit daily written reports to the Quality Control Supervisory Representative reporting the quality of masonry construction. Submit copies of masonry inspector reports.

1.4.5 DETAIL DRAWINGS. Submit detail drawings showing bar splice locations. If the Contractor opts to furnish inch-pound CMU products, drawings showing elevation of walls exposed to view and indicating the location of all cut CMU products shall be submitted for approval.. Bent bars shall be identified on a bending diagram and shall be referenced and located on the drawings. Wall dimensions, bar clearances, and wall openings greater than one masonry unit in area shall be shown. No approval will be given to the shop drawings until the Contractor certifies that all openings, including those for mechanical and electrical service, are shown. If, during construction, additional masonry openings are required, the approved shop drawings shall be resubmitted with the additional openings shown along with the proposed changes. Location of these additional openings shall be clearly highlighted. The minimum scale for wall elevations shall be 1 to 50 ¼ inch per foot. Reinforcement bending details shall conform to the requirements of ACI SP-66. Submit drawings including plans, elevations, and details of wall reinforcement; details of reinforcing bars at corners and wall intersections; offsets; tops, bottoms, and ends of walls; control and expansion joints; lintels; and wall openings.

1.5 DELIVERY, STORAGE, AND HANDLING. Materials shall be delivered, stored, handled, and protected to avoid chipping, breakage, and contact with soil or contaminating material. Store and prepare materials in already disturbed areas to minimize project site disturbance and size of project site.

1.5.1 MASONRY UNITS. Cover and protect moisture-controlled concrete masonry units and cementitious materials from precipitation. Conform to all handling and storage requirements of ASTM C90. Mark prefabricated lintels on top sides to show either the lintel schedule number or the number and size of top and bottom bars.

1.5.2 REINFORCEMENT, ANCHORS, AND TIES. Steel reinforcing bars, coated anchors, ties, and joint reinforcement shall be stored above the ground. Steel reinforcing bars and uncoated ties shall be free of loose mill scale and rust.

1.5.3 CEMENTITIOUS MATERIALS, SAND AND AGGREGATES. Cementitious and other packaged materials shall be delivered in unopened containers, plainly marked and labeled with manufacturers' names and brands. Cementitious material shall be stored in dry, weathertight enclosures or be completely covered. Cement shall be handled in a manner that will prevent the inclusion of foreign materials and damage by water or dampness. Store sand and aggregates in a manner to prevent contamination or segregation.

1.6 PROJECT/SITE CONDITIONS. Conform to ACI 530/530.1 for hot and cold weather masonry erection.

1.6.1 HOT WEATHER INSTALLATION. Take the following precautions if masonry is erected when the ambient air temperature is more than 37 degrees C 99 degrees F in the shade and the relative humidity is less than 50 percent or the ambient air temperature exceeds 32 degrees C 90 degrees F and the wind velocity is more than 13 km/h 8 mph. All masonry materials shall be shaded from direct sunlight; mortar beds shall be spread no more than 1.2 m 4 feet ahead of masonry; masonry units shall be set within one minute of spreading mortar; and after erection, masonry shall be protected from direct exposure to wind and sun for 48 hours.

1.6.2 COLD WEATHER INSTALLATION. Before erecting masonry when ambient temperature or mean daily air temperature falls below 4 degrees C 40 degrees F or temperature of masonry units is below 4 degrees C 40 degrees F, submit a written statement of proposed cold weather construction procedures for approval. Take the additional following precautions if masonry is erected in cold weather: [_____]

PART 2 PRODUCTS

2.1 GENERAL REQUIREMENTS. The source of materials which will affect the appearance of the finished work shall not be changed after the work has started except with Owner's approval. Submit sample of colored mortar with applicable masonry unit and color samples of three stretcher units and one unit for each type of special shape. Units shall show the full range of color and texture. Submit test reports from an approved independent laboratory. Test reports on a previously tested material shall be certified as the same as that proposed for use in this project. Submit certificates of compliance stating that the materials meet the specified requirements.

2.2 CLAY OR SHALE BRICK. Submit brick samples as specified. Color range and texture of clay or shale brick shall be as indicated and shall conform to the approved sample. Brick shall conform to ASTM C62; Grade SW shall be used for brick in contact with earth or grade and for [the first six exterior courses above grade] [all exterior work] and for all nonvertical surfaces. Grade SW or MW shall be used in other brickwork. Average dimensions of brick shall be 90 mm thick, 57 mm high, and 190 mm long (standard) 3-5/8 inches thick, 2-1/4 inches high, and 8 inches long (standard) or 4 inches thick, 2-2/3 inches high, and 8 inches long (nominal), subject to the tolerances specified in ASTM C62. Brick shall be tested for efflorescence. Clay or shale brick units shall be delivered factory-blended to provide a uniform appearance and color range in the completed wall. [Clay units shall contain a minimum of [5] [10] [_____] percent post-consumer recycled content, or a minimum of [20] [40] [_____] percent post-industrial recycled content.] [See Section 01 33 29 LEED(tm) DOCUMENTATION for cumulative total recycled content requirements. Clay units may contain post-consumer or post-industrial recycled content.]

2.2.1 SOLID CLAY OR SHALE BRICK. Solid clay or shale brick shall conform to [ASTM C62] [ASTM C216, Type [FBS] [FBA] [FBX]]. Brick size shall be modular and the nominal size of the brick used shall be 92 mm 3-5/8 inches thick, 57 mm 2-1/4 inches high, and 200 mm 8 inches long (nominal) or 100 mm thick, 68 mm high and 200 mm long (nominal) 4 inches thick, 2-2/3 inches high and 8 inches long (nominal). Minimum compressive strength of the brick shall be [_____] MPa psi.

2.2.2 HOLLOW CLAY OR SHALE BRICK. Hollow clay or shale brick shall conform to ASTM C652, Type [HBS] [HBX] [HBA] [HBB]. Brick size shall be modular and the nominal size of the brick used shall be [_____] mm inches thick, [_____] mm inches high, and [_____] mm inches long. Where vertical reinforcement is shown in hollow brick, the minimum cell dimension shall be 64 mm 2-1/2 inches and the units shall be designed to provide precise vertical alignment of the cells. Minimum compressive strength of the brick shall be [_____] MPa psi.

2.2.3 SAND-LIME BRICK. ASTM C73, Grade SW, approximately 92 mm thick, 57 mm high, 200 mm long (nominal) 3 5/8 inches thick, 2 1/4 inches high, and 8 inches long (nominal) or nominal modular, with smooth surfaces and natural color.

2.2.4 REFRACTORY BRICK. ASTM C27, low-duty type, [_____] mm inches thick, [_____] mm inches high, and [_____] mm inches long.

2.2.5 CLOSURE OR UTILITY BRICK. ASTM C216, Grade SW, Type FBS, [92 mm thick, 92 mm high, and 200 mm long (closure) 3 5/8 inches thick, 3 5/8 inches high, and 8 inches long (closure)] [or] [nominally 100 mm thick, 100 mm high, and 305 mm long (utility) 4 inches thick, 4 inches high, and 12 inches long (utility)]. [Closure] [or] [Utility] brick may be used at the option of the Contractor, provided that changes necessitated by the use of such brick shall be the responsibility of the Contractor. Color, texture, and range of brick shall match the brick [on display at [_____]] [indicated].

2.2.6 ADOBE BRICK. [_____] mm inches thick, [_____] mm inches high, and [_____] mm inches long.

2.2.6.1 TRADITIONAL ADOBE. Protect traditional adobe from water penetration by the application of adobe mud plaster, lime plaster, cement- or lime-cement stucco and wide roof overhangs.

2.2.6.2 SEMI-STABILIZED ADOBE. Semi-stabilized adobe shall contain 3 percent asphalt emulsion or portland cement by weight, or as prescribed by local soil conditions and codes. Protect from water penetration by the application of adobe mud plaster, lime plaster, cement- or lime-cement stucco and wide eaves.

2.2.6.3 FULLY STABILIZED ADOBE. Fully stabilized adobe shall contain 5 percent asphalt emulsion or portland cement by weight, or as prescribed by local soil conditions and codes.

2.3 CONCRETE BRICK. Concrete brick shall conform to ASTM C55, Grade [N] [S]. Concrete brick may be used where necessary for filling out in concrete masonry unit construction. Submit samples as specified.

2.4 SALVAGED BRICK. Use [lead-free] salvaged bricks and other masonry units in place of new bricks or masonry units as indicated. [Bricks salvaged from foundries or industrial buildings shall be washed with appropriate metal-dust removing cleaner.] When using salvaged brick, select exterior face bricks from salvaged exterior face bricks. Bricks shall meet standards of new bricks otherwise used in application, and shall be cleaned of all mortar prior to use. Place exterior face towards the exterior. Submit documentation certifying products are from salvaged/recovered sources. Indicate relative dollar value of salvaged content products to total dollar value of products included in project.

2.5 CONCRETE MASONRY UNITS (CMU). Submit samples and certificates as specified. Cement shall have a low alkali content and be of one brand. [Units shall contain a minimum of [5][10] [_____] percent post-consumer recycled content, or a minimum of [20][40] [_____] percent post-industrial recycled content.][See Section 01

33 29 LEED(tm) DOCUMENTATION for cumulative total recycled content requirements. Units may contain post-consumer or post-industrial recycled content.] Units shall be of modular dimensions and air, water, or steam cured. [Surfaces of units which are to be plastered or stuccoed shall be sufficiently rough to provide bond]; [elsewhere,] [exposed surfaces of units shall be smooth and of uniform texture]. [Exterior concrete masonry units shall have water-repellant admixture added during manufacture.]

a. Hollow Load-Bearing Units: ASTM C90, made with lightweight [or medium weight] [or normal weight] aggregate. Provide load-bearing units for exterior walls, foundation walls, load-bearing walls, and shear walls.

b. Hollow Non-Load-Bearing Units: ASTM C129, made with lightweight [or medium weight] [or normal weight] aggregate. Load-bearing units may be provided in lieu of non-load-bearing units.

c. Solid Load-Bearing Units: ASTM C90, lightweight [or medium weight] [or normal weight] units. Provide solid units [for masonry bearing under structural framing members] [as indicated].

2.5.1 AGGREGATES. Lightweight aggregates and blends of lightweight and heavier aggregates in proportions used in producing the units, shall comply with the following requirements when tested for stain-producing iron compounds in accordance with ASTM C641: by visual classification method, the iron stain deposited on the filter paper shall not exceed the "light stain" classification. Use industrial waste by-products (air-cooled slag, cinders, or bottom ash), ground waste glass and concrete, granulated slag, and expanded slag in aggregates. Slag shall comply with ASTM C989/C989M; Grade [80] [100] [120].

2.5.2 KINDS AND SHAPES. Units shall be modular in size and shall include closer, jamb, header, lintel, and bond beam units and special shapes and sizes to complete the work as indicated. In exposed interior masonry surfaces, units having a bullnose shall be used for vertical external corners except at door, window, and louver jambs. Radius of the bullnose shall be 25 mm 1 inch. Units used in exposed masonry surfaces in any one building shall have a uniform fine to medium texture and a uniform color.

2.5.2.1 ARCHITECTURAL UNITS. Units shall have patterned face shell. Face shell pattern shall be [fluted] [vertical scored] [split ribbed] [_____]. Units shall be integrally colored during manufacture. Color shall be [_____]. Patterned face shell shall be properly aligned in the completed wall.

2.5.2.2 PATTERNED, DECORATIVE SCREEN UNITS. Patterned, decorative screen units shall conform to the applicable requirements of [ASTM C90] [ASTM C129]. Units shall have uniform through-the-wall pattern, color, and texture.

2.5.3 FIRE-RATED CMU. Concrete masonry units used in fire-rated construction shown on the drawings shall be of minimum equivalent thickness for the fire rating indicated and the corresponding type of aggregates indicated in TABLE I. Units containing more than one of the aggregates listed in TABLE I will be rated on the aggregate requiring the greater minimum equivalent thickness to produce the required fire rating. Construction shall conform to ASTM E119.

TABLE I

FIRE-RATED CONCRETE MASONRY UNITS

See note (a) in Table III

Aggregate Type	Minimum equivalent thickness in mm inches for fire rating of:		
	4 hours	3 hours	2 hours
Pumice	1204.7	1004.0	753.0
Expanded slag	1305.0	1104.2	853.3
Expanded clay, shale, or slate	1455.7	1204.8	953.7
Limestone, scoria, cinders or unexpanded slag	1505.9	1305.0	1004.0
Calcareous gravel	1606.2	1355.3	1054.2
Siliceous gravel	1706.7	1455.7	1154.5

Minimum equivalent thickness shall equal net volume as determined in conformance with ASTM C140/C140M divided by the product of the actual length and height of the face shell of the unit in mm inches. Where walls are to receive plaster or be faced with brick, or otherwise form an assembly; the thickness of plaster or brick or other material in the assembly will be included in determining the equivalent thickness. Submit calculation results.

2.6 COMPRESSED EARTH BLOCK. Earth may be stabilized by adding 2 to 5 percent portland cement by weight for semi-stabilized block, and 5 to 10 percent for fully stabilized block; use of other stabilizing admixtures, and their proportions, shall be prescribed by local soil conditions and codes. [250][_____] mm [10][_____] inches thick, [100][_____] mm [4][_____] inches high, and [356][_____] mm [14][_____] inches long, 17 to 19 kg 38 to 40 pounds, using an average of 7,585 kPa 1,100 psi for unstabilized and 20,685 kPa 3,000 psi for stabilized.

2.7 PRECAST CONCRETE ITEMS. Trim, lintels, copings, splashblocks and door sills shall be factory-made units from a plant regularly engaged in producing precast concrete units. Unless otherwise indicated, concrete shall be [28] [20] MPa [4,000] [3000] psi minimum conforming to Section CAST-IN-PLACE CONCRETE using 13 mm 1/2 inch to No. 4 nominal-size coarse aggregate, and minimum reinforcement shall be the reinforcement required for handling of the units. Clearance of 19 mm 3/4 inch shall be maintained between reinforcement and faces of units. Unless precast-concrete items have been subjected during manufacture to saturated-steam pressure of at least 827 kPa 120 psi for at least 5 hours, the items, after casting, shall be either damp-cured for 24 hours or steam-cured and shall then be aged under cover for 28 days or longer. Cast-concrete members weighing over 35 kg 80 pounds shall have built-in loops of galvanized wire or other approved provisions for lifting and anchoring. Units shall have beds and joints at right angles to the face, with sharp true arises and shall be cast with drip grooves on the underside where units overhang walls. Exposed-to-view surfaces shall be free of surface voids, spalls, cracks, and chipped or broken edges. Precast units exposed-to-view shall be of uniform appearance and color. Unless otherwise specified, units shall have a smooth dense finish. Prior to use, each item shall be wetted and inspected for crazing. Items showing evidence of dusting, spalling, crazing, or having surfaces treated with a protective coating will be rejected. Submit specified factory certificates.

2.7.1 LINTELS. Precast lintels, unless otherwise shown, shall be of a thickness equal to the wall and reinforced with two No. 4 bars for the full length. Top of lintels shall be labeled "TOP" or otherwise identified and each lintel shall be clearly marked to show location in the structure. In reinforced masonry, lintels shall conform to ACI 318M ACI 318 for flexural and shear strength and shall have at least 200 mm 8 inches bearing at each end. Concrete shall have a minimum 28 day compressive strength of [_____] MPa psi using 13 mm 1/2 inch to No. 4 nominal-size coarse aggregate. Reinforcement shall conform to ASTM A615/A615M Grade 400 MPa 60,000 psi. Limit lintel deflection due to dead plus live load to L/600 or 7 mm 0.3 inches. Provide top and bottom bars for lintels over 900 mm 36 inches in length.

2.7.2 SILLS AND COPINGS. Sills and copings shall be cast with washes. Sills for windows having mullions shall be cast in sections with head joints at mullions and a 6 mm 1/4 inch allowance for mortar joints. The ends of sills, except a 19 mm 3/4 inch wide margin at exposed surfaces, shall be roughened for bond. Treads of door sills shall have rounded nosings. [Reinforce sills with not less than two No. 15 No. 4 bars.]

2.7.3 SPLASH BLOCKS. Splash blocks shall be as detailed. Reinforcement shall be the manufacturer's standard.

2.7.4 FLUE LININGS AND THIMBLES. ASTM C315, free from fractures. Sizes and shapes shall be as indicated.

2.8 STONE ITEMS. \Submit samples as specified. Stone for trim, sills, lintels, and copings shall be limestone, sandstone, or granite, and shall be cut to the design shown. Sandstone shall be standard grade, buff, gray, or buff brown, with a smooth finish free from clay pits and tool marks. Granite shall be a good commercial grade building granite of medium or moderately coarse grain, and a light or medium gray or light pink color, with a smooth machine finish on washes, 4-cut finish on treads, and 6-cut or equivalent machine finish on other exposed surfaces. Limestone shall be standard buff limestone with a smooth machine finish free from tool marks. Lintels, except when supported by a steel member, shall be 100 mm 4 inches or more thick from face to back edge and of the depth required to support the masonry over the opening. Stone shall have beds and joints at right angles to the face, with sharp, true arises. Copings and sills shall be provided with washes, and where overhanging the walls, shall have drips cut on the underside.

2.9 MORTAR FOR STRUCTURAL MASONRY

ASTM C270 TABLE 2 Property Specification Requirements (For laboratory prepared mortar only)					
Mortar	Type	Average Compressive Strength at 28 Days Min. MPa (psi)	Water Retention Min. Percent	Air Content Max. Percent	Aggregate Ratio (Measured in Damp, Loose Condition)
Cement-Lime	M S N O	17 (2500) 12 (1800) 5 (750) 2 (350)	75 75 75 75	12 12 14* 14*	Not less than 2-1/4 and not more than 3-1/2 times the
Masonry cement	M S N O	17 (2500) 12 (1800) 5 (750) 2 (350)	75 75 75 75	** ** ** **	Separate volumes of cementitious materials
* When structural reinforcement is incorporated in cement-lime mortar, the maximum air content shall be 12 percent.					
** When structural reinforcement is incorporated in masonry cement mortar, the maximum air content shall be 18 percent.					
Type N should be used only for non-load-bearing walls. Approximate the f'm of the unit masonry. Mortars should be slightly weaker than masonry units so that cracking will occur in joints where easy to repair.					

ASTM C270, Type [M] [N] [S]. Strength (f'm) as indicated. Test in accordance with ASTM C780. [Use Type [I] [II] [III] portland cement.] [Use Type [IS] [IP] [I(PM)] blended hydraulic cement.] [Use Masonry cement.] Do not use admixtures containing chlorides. When structural reinforcement is incorporated, maximum air-content shall be 12 percent in cement-lime mortar and 18 percent in masonry cement mortar. Use up to 40 percent Class F fly ash with type IP cement in cement-lime mortar. Fly ash shall comply with ASTM C593.

2.10 MASONRY MORTAR. Type M mortar shall conform to ASTM C270 and shall be used for foundation walls [, basement walls,] [and piers]. Mortar Type [S] [N] shall conform to the proportion specification of ASTM C270 except Type S cement-lime mortar proportions shall be 1 part cement, 1/2 part lime and 4-1/2 parts aggregate; Type N cement-lime mortar proportions shall be 1 part cement, 1 part lime and 6 parts aggregate. Type N or S mortar shall be used for non-load-bearing, non-shear-wall interior masonry; [approved commercial fire clay mortar or refractory cement (calcium-aluminate) mortar for fire brick and flue liners;] and Type S for remaining masonry work; except where higher compressive strength is indicated on structural drawings. When masonry cement ASTM C91/C91M is used the maximum air content shall be limited to 12 percent and performance equal to cement-lime mortar shall be verified. Verification of masonry cement performance shall be based on ASTM C780 and ASTM C1072. Pointing mortar in showers and kitchens shall contain ammonium stearate, or aluminum tri-stearate, or calcium stearate in an amount equal to 3 percent by weight of cement used. Cement shall have a low alkali content and be of one brand. Aggregates shall be from one source.

2.10.1 ADMIXTURES FOR MASONRY MORTAR. In cold weather, a non-chloride based accelerating admixture may be used subject to approval. Accelerating admixture shall be non-corrosive, shall contain less than 0.2 percent chlorides, and shall conform to ASTM C494/C494M, Type C. Submit the required certifications.

2.10.2 COLORED MORTAR. Mortar coloring shall be added to the mortar used for exposed masonry surfaces to produce a uniform color matching [_____]. Quantity of pigment to cementitious content of the masonry cement shall not exceed [5][_____] by weight; carbon black shall not exceed [1][_____] percent by weight. Quantity of pigment to cementitious content of cement-lime mix shall not exceed [10][_____] percent by weight, carbon black no more than [2][_____] percent by weight. Mortar coloring shall be chemically inert, of finely ground limeproof pigment, and furnished in accurately pre-measured and packaged units that can be added to a measured amount of cement. Compressive strength of colored mortar shall equal [_____].

2.10.3 HYDRATED LIME AND ALTERNATES. Hydrated lime shall conform to ASTM C207, Type [S][SA].

2.10.4 CEMENT. Portland cement shall conform to ASTM C150/C150M, Type I,[IA,] II,[IIA,] or III[, IIIA]. Masonry cement shall conform to ASTM C91/C91M, Type [N][S][M].

Containers shall bear complete instructions for proportioning and mixing to obtain the required types of mortar. Incorporate to the maximum extent, without conflicting with other requirements of this section, up to 40 percent fly ash, up to 70 percent slag, up to 10 percent cenospheres, and up to 10 percent silica fume. When masonry cement is used, submit the manufacturer's printed instructions on proportions of water and aggregates and on mixing to obtain the type of mortar required. Additives shall conform to requirements in Section [CAST-IN-PLACE CONCRETE].

2.10.5 PRE-MIXED MORTAR. Pre-mixed mortar shall conform to ASTM C1142, Type [RN] [RS] [RM]. Submit pre-mixed mortar composition.

2.10.6 SAND AND WATER. Sand shall conform to ASTM C144. Water shall be clean, potable, and free from substances which could adversely affect the mortar.

2.11 WATER-REPELLANT ADMIXTURE. Polymeric type formulated to reduce porosity and water penetration and water absorption of the mortar and masonry units[required to provide for the exterior single-wythe masonry wall water penetration resistance indicated in Paragraph SINGLE-WYTHE MASONRY WALL WATER PENETRATION TEST].

2.12 GROUT AND READY-MIXED GROUT. Grout shall conform to ASTM C476, [fine] [coarse]. Cement used in grout shall have a low alkali content. Grout slump shall be between 200 and [250] [280] mm 8 and [10] [11] inches. Minimum grout strength shall be 14 MPa 2000 psi in 28 days, as tested by ASTM C1019. Use grout subject to the limitations of Table III. Do not change proportions and do not use materials with different physical or chemical characteristics in grout for the work unless additional evidence is furnished that the grout meets the specified requirements. Ready-Mixed grout shall conform to ASTM C94/C94M.

2.12.1 ADMIXTURES FOR GROUT. In cold weather, a non-chloride based accelerating admixture may be used subject to approval; accelerating admixture shall be non-corrosive, shall contain less than 0.2 percent chlorides, and shall conform to ASTM C494/C494M, Type C. In general, air-entrainment, anti-freeze or chloride admixtures shall not be used except as approved by the Owner. Submit required certifications.

2.12.2 GROUT BARRIERS. Grout barriers for vertical cores shall consist of fine mesh wire, fiberglass, or expanded metal.

2.13 ANCHORS, TIES, AND BAR POSITIONERS. Anchors and ties shall be fabricated without drips or crimps and shall be zinc-coated in accordance with ASTM A153/A153M, Class B-2. Steel wire used for anchors and ties shall be fabricated from steel wire conforming to ASTM A82/A82M. Wire ties or anchors in exterior walls shall conform to ASTM A641/A641M. Joint reinforcement in interior walls, and in exterior or interior walls exposed to moist environment shall conform to ASTM A641/A641M; coordinate with paragraph JOINT REINFORCEMENT below. Anchors and ties shall be sized to provide a minimum of 16 mm 5/8 inch mortar cover from either face. Submit two anchors, ties and bar positioners of each type used, as samples.

2.13.1 WIRE MESH TIES. Wire mesh for tying 100 mm 4 inch thick concrete masonry unit partitions to other intersecting masonry partitions shall be 13 mm 1/2 inch mesh of minimum 16 gauge 16 gauge steel wire. Minimum lengths shall be not less than 300 mm 12 inches.

2.13.2 WALL TIES. Provide wall ties rectangular-shaped or Z-shaped fabricated of 5 mm 3/16 inch diameter zinc-coated steel wire. Rectangular wall ties shall be no less than 100 mm 4 inches wide. Wall ties may also be of a continuous type conforming to paragraph JOINT REINFORCEMENT. Adjustable type wall ties, if approved for use, shall consist of two essentially U-shaped elements fabricated of 5 mm 3/16 inch diameter zinc-coated steel wire. Adjustable ties shall be of the double pintle to eye type and shall allow a maximum of 13 mm 1/2 inch eccentricity between each element of the tie. Play between pintle and eye opening shall be not more than 2 mm 1/16 inch. The pintle and eye elements shall be formed so that both can be in the same plane.

2.13.3 DOVETAIL ANCHORS. Provide dovetail anchors of the flexible wire type, 5 mm 3/16 inch diameter zinc-coated steel wire, triangular shaped, and attached to a 12 gauge 12 gauge or heavier steel dovetail section. Use these anchors for anchorage of veneer wythes or composite-wall facings extending over the face of concrete columns, beams, or walls. Fill cells within vertical planes of these anchors solid with grout for full height of walls or partitions, or solid units may be used. Dovetail slots are specified in Section [CAST-IN-PLACE CONCRETE].

2.13.4 ADJUSTABLE ANCHORS. Adjustable anchors shall be 5 mm 3/16 inch diameter steel wire, triangular-shaped. Anchors attached to steel shall be 8 mm 5/16 inch diameter steel bars placed to provide 2 mm 1/16 inch play between flexible anchors and structural steel members. Spacers shall be welded to rods and columns. Equivalent welded-on steel anchor rods or shapes standard with the flexible-anchor manufacturer may be furnished when approved. Welds shall be cleaned and given one coat of zinc-rich touch up paint.

2.13.5 BAR POSITIONERS. Bar positioners, used to prevent displacement of reinforcing bars during the course of construction, shall be factory fabricated from 9 gauge steel wire or equivalent, and coated with a hot-dip galvanized finish. Not more than one wire shall cross the cell. Telescoping bar positioner shall be manufactured from AISI 1065 spring steel and coated in accordance with ASTM B633.

2.14 JOINT REINFORCEMENT

	Long. wires	Cross wires
Standard	3.8 mm (9 gauge (0.1483 inch))	3.8 mm
Heavy Duty	4.8 mm (3/16 inch (0.1875 inch))	3.8 mm (9 gauge (0.1483 inch))
Extra Heavy Duty	4.8 mm (3/16 inch (0.1875 inch))	4.8 mm (3/16 inch (0.1875 inch))

Reinforcement made with 4.2 mm (8 gauge) wire is considered special and is not available from all manufacturers.

Joint reinforcement shall be factory fabricated from steel wire conforming to ASTM A82/A82M, welded construction. Tack welding will not be acceptable in reinforcement used for wall ties. Wire shall have zinc coating conforming to ASTM A153/A153M, Class B-2. All wires shall be a minimum of [9] [_____] gauge. Reinforcement shall be ladder type design, having one longitudinal wire in the mortar bed of each face shell for hollow units and one wire for solid units. Joint reinforcement shall be placed a minimum of 16 mm 5/8 inch cover from either face. The distance between crosswires shall not exceed 400 mm 16 inches. Joint reinforcement for straight runs shall be furnished in flat sections not less than 3 m 10 feet long. Joint reinforcement shall be provided with factory formed corners and intersections. If approved for use, joint reinforcement may be furnished with adjustable wall tie features. Submit one piece of each type used, including corner and wall intersection pieces, showing at least two cross wires.

2.15 REINFORCING STEEL BARS AND RODS. Reinforcing steel bars and rods shall conform to ASTM A615/A615M, Grade 60.

2.16 CONTROL JOINT KEYS. Control joint keys shall be a factory fabricated solid section of natural or synthetic rubber (or combination thereof) conforming to ASTM D2000 or polyvinyl chloride conforming to ASTM D2287. The material shall be resistant to oils and solvents. The control joint key shall be provided with a solid shear section not less than 16 mm 5/8 inch thick and 10 mm 3/8 inch thick flanges, with a tolerance of plus or minus 2 mm 1/16 inch. The control joint key shall fit neatly, but without forcing, in masonry unit jamb sash grooves. The control joint key shall be flexible at a temperature of minus 34 degrees C minus 30 degrees F after five hours exposure, and shall have a durometer hardness of not less than 70 when tested in accordance with ASTM D2240.

2.17 RIGID BOARD-TYPE INSULATION. Provide rigid board-type insulation as specified in Section BOARD AND BLOCK INSULATION. Submit one piece of each type used, including corner and wall intersection pieces, showing at least two cross wires. Submit certificate attesting that the polyurethane or polyisocyanurate insulation furnished for the project contains recovered material, and showing an estimated percent of such recovered material.

2.18 EXPANSION-JOINT MATERIALS. Backer rod and sealant shall be adequate to accommodate joint compression equal to 50 percent of the width of the joint. The backer rod shall be compressible rod stock of polyethylene foam, polyurethane foam, butyl rubber foam, or other flexible, nonabsorptive material as recommended by the sealant manufacturer. Sealant shall conform to Section JOINT SEALANTS, and shall be penetrating [with a maximum volatile organic compound (VOC) content of 600 grams/liter]. Submit one piece of each type of material used.

2.19 THROUGH WALL FLASHING. Provide Through Wall Flashing as specified in Section FLASHING AND SHEET METAL. Provide one of the following types [except that flashing indicated to terminate in reglets shall be metal or coated-metal flashing] [and] [except that the material shall be one which is not adversely affected by dampproofing material.]

2.19.1 COATED-COPPER FLASHING. 0.2 kg 7 ounce, electrolytic copper sheet, uniformly coated on both sides with acidproof, alkaliproof, elastic bituminous compound. Factory apply coating to a weight of not less than 1.8 kg/square meter 6 ounces/squarefoot (approximately 0.9 kg/square meter 3 ounces/square foot on each side).

2.19.2 COPPER OR STAINLESS STEEL FLASHING. Copper, ASTM B370, minimum 450 kg 16 ounce weight; stainless steel, ASTM A167, Type 301, 302, 304, or 316, 0.4 mm 0.015 inch thick, No. 2D finish. Provide with factory-fabricated deformations that mechanically bond flashing against horizontal movement in all directions. Deformations shall consist of dimples, diagonal corrugations, or a combination of dimples and transverse corrugations.

[**2.19.3 REINFORCED MEMBRANE FLASHING.** Polyester film core with a reinforcing fiberglass scrim bonded to one side. The membrane shall be impervious to moisture, flexible, and not affected by caustic alkalis. The material, after being exposed for not less than 1/2 hour to a temperature of 0 degrees C 32 degrees F, shall show no cracking when, at that temperature, it is bent 180 degrees over a 2 mm 1/16 inch diameter mandrel and then bent at the same point over the same size mandrel in the opposite direction 360 degrees.]

2.20 WEEP HOLE VENTILATORS. Weep hole ventilators shall be prefabricated aluminum, plastic or wood blocking sized to form the proper size opening in head joints. Provide aluminum and plastic inserts with grill or screen-type openings designed to allow the passage of moisture from cavities and to prevent the entrance or insects. Ventilators shall be sized to match modular construction with a standard 10 mm 3/8 inch mortar joint.

PART 3 EXECUTION

3.1 PREPARATION. Prior to start of work, masonry inspector shall verify the applicable conditions as set forth in ACI 530/530.1, inspection. The Owner will serve as inspector or will select a masonry inspector.

3.1.1 PROTECTION. Ice or snow formed on the masonry bed shall be thawed by the application of heat. Heat shall be applied carefully until the top surface of the masonry is dry to the touch. Sections of masonry deemed frozen and damaged shall be removed before continuing construction of those sections.

3.1.1.1 AIR TEMPERATURE 4 TO 0 DEGREES C 40 TO 32 DEGREES F. Heat sand or mixing water to produce mortar temperatures between 4 and 49 degrees C 40 and 120 degrees F

3.1.1.2 AIR TEMPERATURE 0 TO MINUS 4 DEGREES C 32 TO 25 DEGREES F. Heat sand and mixing water to produce mortar temperatures between 4 and 49 degrees C 40 and 120 degrees F. Maintain temperature of mortar on boards above freezing.

3.1.1.3 AIR TEMPERATURE MINUS 4 TO MINUS 7 DEGREES C 25 TO 20 DEGREES F. Heat sand and mixing water to provide mortar temperatures between 4 and 49 degrees C 40 and 120 degrees F. Maintain temperature of mortar on boards above freezing. Use sources of heat on both sides of walls under construction. Employ windbreaks when wind is in excess of 24 km/hour 15 mph.

3.1.1.4 AIR TEMPERATURE MINUS 7 DEGREES C 20 DEGREES F AND BELOW. Heat sand and mixing water to provide mortar temperatures between 4 and 49 degrees C 40 and 120 degrees F. Provide enclosure and auxiliary heat to maintain air temperature above 0 degrees C 32 degrees F. Temperature of units when laid must not be less than minus 7 degrees C 20 degrees F.

3.1.2 COMPLETED MASONRY AND MASONRY NOT BEING WORKED ON

3.1.2.1 MEAN DAILY AIR TEMPERATURE 4 TO 0 DEGREES C 40 TO 32 DEGREES F. Protect masonry from rain or snow for 24 hours by covering with weather-resistive membrane.

3.1.2.2 MEAN DAILY AIR TEMPERATURE 0 TO MINUS 4 DEGREES C 32 TO 25 DEGREES F. Completely cover masonry with weather-resistant membrane for 24 hours.

3.1.2.3 MEAN DAILY AIR TEMPERATURE MINUS 4 TO MINUS 7 DEGREES C 25 TO 20 DEGREES F. Completey cover masonry with insulating blankets or equally protected for 24 hours.

3.1.2.4 MEAN DAILY TEMPERATURE MINUS 7 DEGREES C 20 DEGREES F AND BELOW. Maintain masonry temperature above 0 degrees C 32 degrees F for 24 hours by enclosure and supplementary heat, by electric heating blankets, infrared heat lamps, or other approved methods.

3.1.3 STAINS. Protect exposed surfaces from mortar and other stains. When mortar joints are tooled, remove mortar from exposed surfaces with fiber brushes and wooden paddles. Protect base of walls from splash stains by covering adjacent ground with sand, sawdust, or polyethylene.

3.1.4 LOADS. Do not apply uniform loads for at least 12 hours or concentrated loads for at least 72 hours after masonry is constructed. Provide temporary bracing as required.

3.1.5 SURFACES. Clean surfaces on which masonry is to be placed of laitance, dust, dirt, oil, organic matter, or other foreign materials and slightly roughen to provide a surface texture with a depth of at least 3 mm 1/8 inch. Sandblast, if necessary, to remove laitance from pores and to expose the aggregate.

3.2 LAYING MASONRY UNITS

a. Coordinate masonry work with the work of other trades to accommodate built-in items and to avoid cutting and patching. Masonry units shall be laid in [running] [stacked] [the indicated] bond pattern. Facing courses shall be level with back-up courses, unless the use of adjustable ties has been approved in which case the tolerances shall be plus or minus 13 mm 1/2 inch. Each unit shall be adjusted to its final position while mortar is still soft and plastic.

b. Units that have been disturbed after the mortar has stiffened shall be removed, cleaned, and relaid with fresh mortar. Air spaces, cavities, chases, expansion joints, and spaces to be grouted shall be kept free from mortar and other debris. Units used in exposed masonry surfaces shall be selected from those having the least amount of chipped edges or other imperfections detracting from the appearance of the finished work. Vertical joints shall be kept plumb.

c. Units being laid and surfaces to receive units shall be free of water film and frost. Solid units shall be laid in a nonfurrowed full bed of mortar. Mortar for veneer wythes shall be beveled and sloped toward the center of the wythe from the cavity side. Units shall be shoved into place so that the vertical joints are tight. Vertical joints of brick and the vertical face shells of concrete masonry units, except where indicated at control, expansion, and isolation joints, shall be completely filled with mortar. Mortar will be permitted to protrude up to 13 mm 1/2 inch into the space or cells to be grouted. Means shall be provided to prevent mortar from dropping into the space below.

d. In double wythe construction, the inner wythe may be brought up not more than 400 mm 16 inches ahead of the outer wythe. Collar joints shall be filled with mortar or grout during the laying of the facing wythe, and filling shall not lag the laying of the facing wythe by more than 200 mm 8 inches.

3.2.1 FORMS AND SHORES. Provide bracing and scaffolding as required. Design bracing to resist wind pressure as required by local codes. Forms and shores shall be sufficiently rigid to prevent deflections which may result in cracking or other damage to supported masonry and sufficiently tight to prevent leakage of mortar and grout. Supporting forms and shores shall not be removed in less than 10 days.

3.2.2 REINFORCED CONCRETE MASONRY UNITS WALLS. Where vertical reinforcement occurs, fill cores solid with grout. Lay units in such a manner as to preserve the unobstructed vertical continuity of cores to be filled. Embed the adjacent webs in mortar to prevent leakage of grout. Remove mortar fins protruding from joints before placing grout. Minimum clear dimensions of vertical cores shall be 50 by 75 mm 2 by 3 inches. Position reinforcing accurately as indicated before placing grout. As masonry work progresses, secure vertical reinforcing in place at vertical intervals not to exceed 160 bar diameters. Use puddling rod or vibrator to consolidate the grout. Minimum clear distance between masonry and vertical reinforcement shall be not less than 13 mm 1/2 inch. Unless indicated or specified otherwise, form splices by lapping bars not less than 40 bar diameters and wire tying them together.

3.2.3 CONCRETE MASONRY UNITS. Units in piers, pilasters, columns, starting courses on footings, solid foundation walls, lintels, and beams, and where cells are to be filled with grout shall be full bedded in mortar under both face shells and webs. Other units shall be full bedded under both face shells. Head joints shall be filled solidly with mortar for a distance in from the face of the unit not less than the thickness of the face shell. Foundation walls below grade shall be grouted solid. Jamb units shall be of the shapes and sizes to conform with wall units. Solid units may be incorporated in the masonry work where necessary to fill out at corners, gable slopes, and elsewhere as approved. Double walls shall be stiffened at wall-mounted plumbing fixtures by use of strap anchors, two above each fixture and two below each fixture, located to avoid pipe runs, and extending from center to center of the double wall. Walls and partitions shall be adequately reinforced for support of wall-hung plumbing fixtures when chair carriers are not specified.

3.2.4 CLAY OR SHALE BRICK UNITS. Lay brick facing with the better face exposed. Lay brick in running bond with each course bonded at corners, unless otherwise indicated. Lay molded brick with the frog side down. Brick that is cored, recessed, or has other deformations may be used in sills, treads, soldier courses, except where deformations will be exposed to view. [Lay fire brick by dipping each brick in a soft mixture of fire clay and water and then rubbing the brick into place with joints as thin as practicable or provide refractory mortar with joints not more than 10 mm 3/8 inch thick.]

3.2.4.1 WETTING OF UNITS. Wetting of clay, shale brick, or hollow brick units having an initial rate of absorption of more than 0.155 gm per minute per square cm 1 gram per minute per square inch of bed surface shall be in conformance with ASTM C67. The method of wetting shall ensure that each unit is nearly saturated but surface dry when laid. Test clay or shale brick daily on the job, prior to laying, as follows: Using a wax pencil, draw a circle the size of a quarter on five randomly selected bricks. Apply 20 drops of water with a medicine dropper to the surface within the circle on each brick. If the average time that the water is completely absorbed in the five bricks is less than 1-1/2 minutes, wet bricks represented by the five bricks tested.

3.2.4.2 SOLID UNITS. Completely fill bed, head, and collar joints with mortar.

3.2.4.3 HOLLOW UNITS. Lay hollow units as specified for concrete masonry units.

3.2.4.4 BRICK-FACED WALLS. For brick-faced walls [bond the two wythes in every sixth brick course with continuous horizontal joint reinforcement.] [bond brick in the pattern as indicated on the drawings.] Provide additional bonding ties spaced not more than one meter 3 feet apart around the perimeter of and within 300 mm 12 inches of all openings.

3.2.4.4.1 COLLAR JOINTS. Fill collar joints solid with mortar as each course of brick is laid. Do not disturb units in place.

© J. Paul Guyer 2014

3.2.4.4.2 BRICK SILLS. Lay brick on edge, slope, and project not less than 13 mm 1/2 inch beyond the face of the wall to form a wash and drip. Fill all joints solidly with mortar and tool.

3.2.4.5 CAVITY WALLS. Provide a continuous cavity as indicated. Securely tie the two wythes together with horizontal joint reinforcement. Bevel mortar beds away from cavity to prevent projection into cavity when bricks are shoved in place. Keep cavities clear and clean of mortar droppings. [At the bottom of cavity walls, in the course immediately above the through-wall flashing, temporarily omit one brick every 1200 mm 4 feet. With a hose and clean water, wash all mortar droppings and debris out of the cavity through the temporary openings at least twice each day masonry is laid, and more often when required to keep the cavities clean. Fill in the openings with bricks and mortar after the wall is complete and the cavity has been inspected and found clean.] Provide weep holes of open head joints spaced 600 mm 24 inches o.c. [wherever the cavity is interrupted] [at base of wall and vertical obstructions (e.g. lintels)]. [Cavity face of interior wythe shall be dampproofed in accordance with Section BITUMINOUS DAMPPROOFING.]

3.2.4.6 REINFORCED BRICK WALLS. Provide two wythes of brick separated by a [_____] mm inch wide continuous space filled with [grout] [bricks "floated" in grout] and reinforced as indicated. Bevel mortar beds away from grout space to prevent projection into grout space when bricks are shoved in place. Deeply furrowed bed joints will not be permitted. Lay exterior wythe of brick to the height of each grout pour in advance of interior wythe. Clean grout space and set reinforcing before laying interior wythe. Provide metal ties to prevent spreading of the wythes and to maintain vertical alignment of walls. Position reinforcing as indicated. Wire vertical reinforcing securely in position as the brickwork progresses. Use puddling rod or vibrator to consolidate the grout. The minimum clear distance between parallel bars shall be the nominal diameter of the bars; the minimum clear distance between masonry and reinforcing shall be 6 mm 1/4 inch.

Unless indicated or specified otherwise, form splices by lapping bars not less than 40 bar diameters and wire tying them together. Stagger splices in adjacent horizontal bars.

3.2.4.7 CHIMNEYS. Construct chimneys of brick with clay flue linings of the sizes indicated. Extend flue linings from 300 mm 12 inches below the smoke inlet to 100 mm 4 inches above the chimney cap. Place thimbles as indicated, flush with inside of or up to 25 mm one inch into the flue lining. Set linings in fire clay mortar or refractory mortar and fill and smooth the joints on the inside. Set each section of flue lining before surrounding brickwork reaches top of flue lining section below. Build brickwork around lining, and [fill the space] [leave a 25 mm one inch airspace] between lining and brickwork [with grout]. [Seal top of airspace before installing chimney cap.] Do not cut linings after they are installed in chimney. Unless indicated otherwise, provide a chimney cap of air-entrained concrete. Slope cap to a minimum edge thickness of 50 mm 2 inches and reinforce with two rings of No. 3 gage galvanized steel wire.

3.2.4.8 BRICK VENEER. Provide a continuous cavity as indicated. Install brick veneer after sheathing, masonry anchors, and flashing have been installed to the cold-formed steel framing system. Care shall be provided to avoid damaging the moisture barrier. Damaged moisture barrier and flashing shall be repaired or replaced before brick veneer is installed. Means shall be provided to keep cavities clean and clear of mortar droppings.

3.2.5 TOLERANCES. Lay masonry plumb, true to line, with courses level. Keep bond pattern plumb throughout. Square corners unless noted otherwise. Except for walls constructed of prefaced concrete masonry units, lay masonry within the following tolerances (plus or minus unless otherwise noted):

TABLE II TOLERANCES	
Variation from the plumb in the lines and surfaces of columns, walls and arises	
In adjacent masonry units	3 mm 1/8 inch
In 3 m 10 feet	6 mm 1/4 inch
In 6 m 20 feet	10 mm 3/8 inch
In 12 m 40 feet or more	13 mm 1/2 inch
Variations from the plumb for external corners, expansion joints, and other conspicuous lines	
In 6 m 20 feet	6 mm 1/4 inch
In 12 m 40 feet or more	13 mm 1/2 inch
Variations from the level for exposed lintels, sills, parapets, horizontal grooves, and other conspicuous lines	
In 6 m 20 feet	6 mm 1/4 inch
In 12 m 40 feet or more	13 mm 1/2 inch
Variation from level for bed joints and top surfaces of bearing walls	
In 3 m 10 feet	6 mm 1/4 inch
In 12 m 40 feet or more	13 mm 1/2 inch
Variations from horizontal lines	
In 3 m 10 feet	6 mm 1/4 inch
In 6 m 20 feet	10 mm 3/8 inch
In 12 m 40 feet or more	13 mm 1/2 inch
Variations in cross sectional dimensions of columns and in thickness of walls	
Minus	6 mm 1/4 inch
Plus	13 mm 1/2 inch

3.2.6 CUTTING AND FITTING. Full units of the proper size shall be used wherever possible, in lieu of cut units. Cutting and fitting, including that required to accommodate the work of others, shall be done by masonry mechanics using power masonry saws. Concrete masonry units may be wet or dry cut. Wet cut units, before being placed in the work, shall be dried to the same surface-dry appearance as uncut units being laid in the wall. Cut edges shall be clean, true and sharp. Openings in the masonry shall be made carefully so that wall plates, cover plates or escutcheons required by the installation will completely conceal the openings and will have bottoms parallel with the masonry bed joints. Reinforced masonry lintels shall be provided above openings over 300 mm 12 inches wide for pipes, ducts, cable trays, and other wall penetrations, unless steel sleeves are used.

3.2.7 JOINTING. Joints shall be tooled when the mortar is thumbprint hard. Horizontal joints shall be tooled last. Joints shall be brushed to remove all loose and excess mortar. Mortar joints shall be finished as follows:

3.2.7.1 FLUSH JOINTS. Joints in concealed masonry surfaces and joints at electrical outlet boxes in wet areas shall be flush cut. Flush cut joints shall be made by cutting off the mortar flush with the face of the wall. Joints in unparged masonry walls below grade shall be pointed tight. Flush joints for architectural units, such as fluted units, shall completely fill both the head and bed joints.

3.2.7.2 TOOLED JOINTS. Joints in exposed exterior and interior masonry surfaces shall be tooled [slightly concave] [_____]. Joints shall be tooled with a jointer slightly larger than the joint width so that complete contact is made along the edges of the unit. Tooling shall be performed so that the mortar is compressed and the joint surface is sealed. Jointer of sufficient length shall be used to obtain a straight and true mortar joint.

3.2.7.3 DOOR AND WINDOW FRAME JOINTS. On the exposed interior side of exterior frames, joints between frames and abutting masonry walls shall be raked to a depth of

10 mm 3/8 inch. On the exterior side of exterior frames, joints between frames and abutting masonry walls shall be raked to a depth of 10 mm 3/8 inch.

3.2.8 JOINT WIDTHS. Joint widths shall be as follows:

3.2.8.1 CONCRETE MASONRY UNITS. Concrete masonry units shall have 10 mm 3/8 inch joints, except for prefaced concrete masonry units.

3.2.8.2 PREFACED CONCRETE MASONRY UNITS. Prefaced concrete masonry units shall have a joint width of 10 mm 3/8 inch wide on unfaced side and not less than 5 mm 3/16 inch nor more than 6 mm 1/4 inch wide on prefaced side.

3.2.8.3 BRICK. Brick joint widths shall be the difference between the actual and nominal dimensions of the brick in either height or length. Brick expansion joint widths shall be as shown.

3.2.9 EMBEDDED ITEMS. Fill spaces around built-in items with mortar. Point openings around flush-mount electrical outlet boxes in wet locations with mortar. Embed anchors, ties, wall plugs, accessories, flashing, pipe sleeves and other items required to be built-in as the masonry work progresses. Fully embed anchors, ties and joint reinforcement in the mortar. Fill cells receiving anchor bolts and cells of the first course below bearing plates with grout.

3.2.10 UNFINISHED WORK. Step back unfinished work for joining with new work. Toothing may be resorted to only when specifically approved. Remove loose mortar and thoroughly clean the exposed joints before laying new work.

3.2.11 MASONRY WALL INTERSECTIONS. Masonry bond each course at corners and elsewhere as shown. Masonry walls shall be anchored or tied together at corners and intersections with bond beam reinforcement and prefabricated corner or tee pieces of joint reinforcement as shown.

3.2.12 PARTITIONS. Partitions shall be continuous from floor to underside of floor or roof deck where shown. Openings in firewalls around joists or other structural members shall be filled as indicated or approved. Where suspended ceilings on both sides of partitions are indicated, the partitions other than those shown to be continuous may be stopped approximately 100 mm 4 inches above the ceiling level. An isolation joint shall be placed in the intersection between partitions and structural or exterior walls as shown. Interior partitions having 100 mm 4 inch nominal thick units shall be tied to intersecting partitions of 100 mm 4 inch units, 125 mm 5 inches into partitions of 150 mm 6 inch units, and 175 7 inches into partitions of 200 mm 8 inch or thicker units. Cells within vertical plane of ties shall be filled solid with grout for full height of partition or solid masonry unit may be used. Interior partitions having masonry walls over 100 mm 4 inches thick shall be tied together with joint reinforcement. Partitions containing joint reinforcement shall be provided with prefabricated pieces at corners and intersections or partitions.

3.3 COMPRESSED EARTH BLOCK. Install according to manufacturer instructions and accepted industry standards.

3.4 ANCHORED VENEER CONSTRUCTION. Completely separate the inner and outer wythes by a continuous airspace as indicated. Lay up both the inner and the outer wythes together except when adjustable joint reinforcement assemblies are approved for use. When both wythes are not brought up together, through-wall flashings shall be protected from damage until they are fully enclosed in the wall. The airspace between the wythes shall be kept clear and free of mortar droppings by temporary wood strips laid on the wall ties and carefully lifted out before placing the next row of ties. A coarse gravel or drainage material shall be placed behind the weep holes in the cavity to a minimum depth of 100 mm 4 inches of coarse aggregate or 250 mm 10 inches of drainage material to keep mortar droppings from plugging the weep holes.

3.5 WEEP HOLES

Wherever through-wall flashing occurs, provide weep holes to drain flashing to exterior at acceptable locations as indicated on drawings. Weep holes shall be [open head joints].[clear round holes not less than 6 mm 1/4 inch in diameter] at 600 mm 24 inches o.c. Weep holes shall be provided not more than 600 mm 24 inches on centers in mortar joints of the exterior wythe above wall flashing, over foundations, bond beams, and any other horizontal interruptions of the cavity. Weep holes shall be perfectly horizontal or slightly canted downward to encourage water drainage outward and not inward. [Weep holes shall be formed by placing short lengths of well-greased No. 10, 8 mm 5/16 inch nominal diameter, braided cotton sash cord in the mortar and withdrawing the cords after the wall has been completed.] [Weep holes shall be constructed using weep hole ventilators.] Other approved methods may be used for providing weep holes. Weep holes shall be kept free of mortar and other obstructions.

3.6 COMPOSITE WALLS. Tie masonry wythes together with joint reinforcement or with unit wall ties. Anchor facing to concrete backing with wire dovetail anchors set in slots built in the face of the concrete as specified in Section CAST-IN-PLACE CONCRETE][03 30 00 CAST-IN-PLACE CONCRETE]. Anchor or tie the facing wythe to the backup at a maximum spacing of 400 mm 16 inches on center vertically and 600 mm 24 inches on center horizontally. Unit ties shall be spaced not over 600 mm 24 inches on centers horizontally, in courses not over 400 mm 16 inches apart vertically, staggered in alternate courses. Ties shall be laid not closer than 16 mm 5/8 inch to either masonry face. Ties shall not extend through control joints. Collar joints between masonry facing and masonry backup shall be filled solidly with grout.

3.7 MORTAR MIX. Mix mortar in a mechanically operated mortar mixer for at least 3 minutes, but not more than 5 minutes. Measure ingredients for mortar by volume. Ingredients not in containers, such as sand, shall be accurately measured by the use of measuring boxes. Mix water with the dry ingredients in sufficient amount to provide a workable mixture which will adhere to the vertical surfaces of masonry units. Retemper mortar that has stiffened because of loss of water through evaporation by adding water to restore the proper consistency and workability. Discard mortar that has reached its initial set or that has not been used within [2.5] [_____] hours after mixing.

3.8 REINFORCING STEEL. Clean reinforcement of loose, flaky rust, scale, grease, mortar, grout, or other coating which might destroy or reduce its bond prior to placing grout. Bars with kinks or bends not shown on the drawings shall not be used. Reinforcement shall be placed prior to grouting. Unless otherwise indicated, vertical wall reinforcement shall extend to within 50 mm 2 inches of tops of walls.

3.8.1 POSITIONING BARS. Vertical bars shall be accurately placed within the cells at the positions maintained between the bars and masonry units. Minimum clearance between parallel bars shall be one diameter of the reinforcement. Vertical reinforcing may be held in place using bar positioners located near the ends of each bar and at intermediate intervals of not more than 192 diameters of the reinforcement. Column and pilaster ties shall be wired in position around the vertical steel. Ties shall be in contact with the vertical reinforcement and shall not be placed in horizontal bed joints.

3.8.2 SPLICES. Bars shall be lapped a minimum of 48 diameters of the reinforcement. Welded or mechanical connections shall develop at least 125 percent of the specified yield strength of the reinforcement.

3.9 JOINT REINFORCEMENT INSTALLATION. Install joint reinforcement at 400 mm 16 inches on center or as indicated. Reinforcement shall be lapped not less than 150 mm 6 inches. Install prefabricated sections at corners and wall intersections. Place the longitudinal wires of joint reinforcement to provide not less than 16 mm 5/8 inch cover to either face of the unit.

3.10 PLACING GROUT. Fill cells containing reinforcing bars with grout. Hollow masonry units in walls or partitions supporting plumbing, heating, or other mechanical fixtures, voids at door and window jambs, and other indicated spaces shall be filled solid with grout. Cells under lintel bearings on each side of openings shall be filled solid with grout for full height of openings. Walls below grade, lintels, and bond beams shall be filled solid with grout. Units other than open end units may require grouting each course to preclude voids in the units. Grout not in place within 1-1/2 hours after water is first added to the batch shall be discarded. Sufficient time shall be allowed between grout lifts to preclude displacement or cracking of face shells of masonry units. If blowouts, flowouts, misalignment, or cracking of face shells should occur during construction, the wall shall be torn down and rebuilt.

3.10.1 VERTICAL GROUT BARRIERS FOR FULLY GROUTED WALLS. Provide grout barriers not more than 10 m 30 feet apart, or as required, to limit the horizontal flow of grout for each pour.

3.10.2 HORIZONTAL GROUT BARRIERS. Embed grout barriers in mortar below cells of hollow units receiving grout.

3.10.3 GROUT HOLES AND CLEANOUTS

3.10.3.1 GROUT HOLES. Provide grouting holes in slabs, spandrel beams, and other in-place overhead construction. Locate holes over vertical reinforcing bars or as required to facilitate grout fill in bond beams. Provide additional openings spaced not more than 400 mm 16 inches on centers where grouting of all hollow unit masonry is indicated. Openings shall not be less than 100 mm 4 inches in diameter or 75 by 100 mm 3 by 4 inches in horizontal dimensions. Upon completion of grouting operations, plug and finish grouting holes to match surrounding surfaces.

3.10.3.2 CLEANOUTS FOR HOLLOW UNIT MASONRY CONSTRUCTION. Provide cleanout holes at the bottom of every pour in cores containing vertical reinforcement when the height of the grout pour exceeds 1.5 m 5 feet. Where all cells are to be grouted, construct cleanout courses using bond beam units in an inverted position to permit cleaning of all cells. Provide cleanout holes at a maximum spacing of 800 mm 32 inches where all cells are to be filled with grout. Establish a new series of cleanouts if grouting operations are stopped for more than 4 hours. Cleanouts shall not be less than 75 by 100 mm 3 by 4 inch openings cut from one face shell. Manufacturer's standard cutout units may be used at the Contractor's option. Cleanout holes shall not be closed until masonry work, reinforcement, and final cleaning of the grout spaces have been completed and inspected. For walls which will be exposed to view, close cleanout holes in an approved manner to match surrounding masonry.

3.10.3.3 CLEANOUTS FOR SOLID UNIT MASONRY CONSTRUCTION. Provide cleanouts for construction of walls consisting of a grout filled cavity between solid masonry wythes at the bottom of every pour by omitting every other masonry unit from one wythe. Establish a new series of cleanouts if grouting operations are stopped for more than 4 hours. Do not plug cleanout holes until masonry work, reinforcement, and final cleaning of the grout spaces have been completed and inspected. For walls which will be exposed to view, close cleanout holes in an approved manner to match surrounding masonry.

3.10.4 GROUTING EQUIPMENT

3.10.4.1 GROUT PUMPS. Pumping through aluminum tubes will not be permitted. Operate pumps to produce a continuous stream of grout without air pockets, segregation, or contamination. Upon completion of each day's pumping, remove waste materials and debris from the equipment, and dispose of outside the masonry.

3.10.4.2 VIBRATORS. Internal vibrators shall maintain a speed of not less than 5,000 impulses per minute when submerged in the grout. Maintain at least one spare vibrator at the site at all times. Apply vibrators at uniformly spaced points not further apart than the visible effectiveness of the machine. Limit duration of vibration to time necessary to produce satisfactory consolidation without causing segregation.

3.10.5 GROUT PLACEMENT. Lay masonry to the top of a pour before placing grout. Do no place grout in two-wythe solid unit masonry cavity until mortar joints have set for at least 3 days during hot weather and 5 days during cold damp weather. Grout shall not be placed in hollow unit masonry until mortar joints have set for at least 24 hours. Grout shall be placed using a hand bucket, concrete hopper, or grout pump to completely fill the grout spaces without segregation of the aggregates. Vibrators shall not be inserted into lower pours that are in a semi-solidified state. The height of grout pours and type of grout used shall be limited by the dimensions of grout spaces as indicated in Table III. Low-lift grout methods may be used on pours up to and including 1.5 m 5 feet in height. High-lift grout methods shall be used on pours exceeding 1.5 m 5 feet in height.

3.10.5.1 LOW-LIFT METHOD. Grout shall be placed at a rate that will not cause displacement of the masonry due to hydrostatic pressure of the grout. Mortar protruding more than 13 mm 1/2 inch into the grout space shall be removed before beginning the grouting operation. Grout pours 300 mm 12 inches or less in height shall be consolidated by mechanical vibration or by puddling. Grout pours over 300 mm 12 inches in height shall be consolidated by mechanical vibration and reconsolidated by mechanical vibration after initial water loss and settlement has occurred. Vibrators shall not be inserted into lower pours that are in a semi-solidified state. Low-lift grout shall be used subject to the limitations of Table III.

3.10.5.2 HIGH-LIFT METHOD. Mortar droppings shall be cleaned from the bottom of the grout space and from reinforcing steel. Mortar protruding more than 6 mm 1/4 inch into the grout space shall be removed by dislodging the projections with a rod or stick as the work progresses. Reinforcing, bolts, and embedded connections shall be rigidly held in position before grouting is started. CMU units shall not be pre-wetted. Grout, from the mixer to the point of deposit in the grout space shall be placed as rapidly as practical by pumping and placing methods which will prevent segregation of the mix and cause a minimum of grout splatter on reinforcing and masonry surfaces not being immediately encased in the grout lift. The individual lifts of grout shall be limited to 1.2 m 4 feet in height. The first lift of grout shall be placed to a uniform height within the pour section and vibrated thoroughly to fill all voids. This first vibration shall follow immediately behind the pouring of the grout using an approved mechanical vibrator. After a waiting period sufficient to permit the grout to become plastic, but before it has taken any set, the succeeding lift shall be poured and vibrated 300 to 450 mm 12 to 18 inches into the preceding lift. If the placing of the succeeding lift is going to be delayed beyond the period of workability of the preceding, each lift shall be reconsolidated by reworking with a second vibrator as soon as the grout has taken its settlement shrinkage. The waiting, pouring, and reconsolidation steps shall be repeated until the top of the pour is reached. The top lift shall be reconsolidated after the required waiting period. The high-lift grouting of any section of wall between vertical grout barriers shall be completed to the top of a pour in one working day unless a new series of cleanout holes is established and the resulting horizontal construction joint cleaned. High-lift grout shall be used subject to the limitations in Table III.

TABLE III
POUR HEIGHT AND TYPE OF GROUT FOR VARIOUS GROUT SPACE DIMENSIONS

Maximum Grout Pour Height in feet (4)	Grout Type	Grouting Procedure	Minimum Dimensions of the Total Clear Areas Within Grout Spaces and Cells in mm inches (1,2)	
			Multiwythe Masonry (3)	Hollow-unit Masonry
0.31	Fine	Low Lift	20 3/4	40 x 50 1-1/2 x 2
1.55	Fine	Low Lift	50 2	50 x 75 2 x 3
2.48	Fine	High Lift	50 2	50 x 75 2 x 3
3.612	Fine	High Lift	65 2-1/2	65 x 75 2-1/2 x 3
7.324	Fine	High Lift	75 3	75 x 75 3 x 3
0.31	Coarse	Low Lift	40 1-1/2	40 x 75 1-1/2 x 3
1.55	Coarse	Low Lift	50 2	65 x 75 2-1/2 x 3
2.48	Coarse	High Lift	50 2	75 x 75 3 x 3

TABLE III
POUR HEIGHT AND TYPE OF GROUT FOR VARIOUS GROUT SPACE DIMENSIONS

			Minimum Dimensions of the Total Clear Areas Within Grout Spaces and Cells in mm inches (1,2)	
Maximum Grout Pour Height m feet (4)	Grout Type	Grouting Procedure	Multiwythe Masonry (3)	Hollow-unit Masonry
3.6 12	Coarse	High Lift	65 2-1/2	75 x 75 3 x 3
7.3 24	Coarse	High Lift	75 3	75 x 100 3 x 4

Notes:
 (1) The actual grout space or cell dimension shall be larger than the sum of the following items:
   (a) The required minimum dimensions of total clear areas given in the table above;
   (b) The width of any mortar projections within the space;
   (c) The horizontal projections of the diameters of the horizontal reinforcing bars within a cross section of the grout space or cell.

 (2) The minimum dimensions of the total clear areas shall be made up of one or more open areas, with at least one area being 20 mm 3/4 inch or greater in width.

 (3) For grouting spaces between masonry wythes.

 (4) Where only cells of hollow masonry units containing reinforcement are grouted, the maximum height of the pour shall not exceed the distance between horizontal bond beams.

3.11 BOND BEAMS. Bond beams shall be filled with grout and reinforced as indicated on the drawings. Grout barriers shall be installed under bond beam units to retain the grout as required. Reinforcement shall be continuous, including around corners, except through control joints or expansion joints, unless otherwise indicated on the drawings. Where splices are required for continuity, reinforcement shall be lapped 48 bar diameters. A minimum clearance of 13 mm 1/2 inch shall be maintained between reinforcement and interior faces of units.

3.12 CONTROL JOINTS. Control joints shall be provided as indicated and shall be constructed by using [mortar to fill the head joint] [special control-joint units] [sash jamb units with control joint key] [open end stretcher units] in accordance with the details shown on the drawings. Sash jamb units shall have a 19 by 19 mm 3/4 by 3/4 inch groove near the center at end of each unit. The vertical mortar joint at control joint locations shall be continuous, including through all bond beams. This shall be accomplished by utilizing half blocks in alternating courses on each side of the joint. The control joint key shall be interrupted in courses containing continuous bond beam steel. In single wythe exterior masonry walls, the exterior control joints shall be raked to a depth of 19 mm 3/4 inch; backer rod and sealant shall be installed in accordance with Section JOINT SEALANTS. Exposed interior control joints shall be raked to a depth of 6 mm 1/4 inch. Concealed control joints shall be flush cut.

3.13 INDICATED JOINTS. [Brick expansion joints][Concrete masonry veneer joints] located, detailed, and constructed as indicated. Keep joints free of mortar and other debris.

3.14 SHELF ANGLES. Adjust shelf angles as required to keep the masonry level and at the proper elevation. Shelf angles shall be galvanized and provided in sections not longer than 3 m 10 feet and installed with a 6 mm 1/4 inch gap between sections. Shelf angles shall be mitered and welded at building corners with each angle not shorter than 1.2 m 4 feet, unless limited by wall configuration.

3.15 LINTELS

3.15.1 MASONRY LINTELS. Construct masonry lintels with lintel units filled solid with grout in all courses and reinforced with a minimum of two No. 4 bars in the bottom course unless otherwise indicated on the drawings. Lintel reinforcement shall extend beyond each side of masonry opening 40 bar diameters or 600 mm 24 inches, whichever is greater. Reinforcing bars shall be supported in place prior to grouting and shall be located 13 mm 1/2 inch above the bottom inside surface of the lintel unit.

3.15.2 PRECAST CONCRETE AND STEEL LINTELS. Construct precast concrete and steel lintels as shown on the drawings. Lintels shall be set in a full bed of mortar with faces plumb and true. Steel and precast lintels shall have a minimum bearing length of 200 mm 8 inches unless otherwise indicated on the drawings.

3.16 SILLS AND COPINGS. Sills and copings shall be set in a full bed of mortar with faces plumb and true.

3.17 ANCHORAGE TO CONCRETE AND STRUCTURAL STEEL

3.17.1 ANCHORAGE TO CONCRETE. Anchorage of masonry to the face of concrete columns, beams, or walls shall be with dovetail anchors spaced not over 400 mm 16 inches on centers vertically and 600 mm 24 inches on center horizontally.

3.17.2 ANCHORAGE TO STRUCTURAL STEEL. Masonry shall be anchored to vertical structural steel framing with adjustable steel wire anchors spaced not over 400 mm 16 inches on centers vertically, and if applicable, not over 600 mm 24 inches on centers horizontally.

3.18 PARGING. The outside face of below-grade exterior concrete-masonry unit walls enclosing usable rooms and spaces, except crawl spaces, shall be parged with type S mortar. Parging shall not be less than 13 mm 1/2 inch thick troweled to a smooth dense surface so as to provide a continuous unbroken shield from top of footings to a line 150 mm 6 inches below adjacent finish grade, unless otherwise indicated. Parging shall be coved at junction of wall and footing. Parging shall be damp-cured for 48 hours or more before backfilling. Parging shall be protected from freezing temperatures until hardened.

3.19 INSULATION. Anchored veneer walls shall be insulated, where shown, by installing board-type insulation on the cavity side of the inner wythe. Board type insulation shall be applied directly to the masonry or thru-wall flashing with adhesive. Insulation shall be neatly fitted between obstructions without impaling of insulation on ties or anchors. The insulation shall be applied in parallel courses with vertical joints breaking midway over the course below and shall be applied in moderate contact with adjoining units without forcing, and shall be cut to fit neatly against adjoining surfaces.

3.20 SPLASH BLOCKS. Locate splash blocks as indicated.

3.21 POINTING AND CLEANING. After mortar joints have attained their initial set, but prior to hardening, completely remove mortar and grout daubs or splashings from masonry-unit surfaces that will be exposed or painted. Before completion of the work, defects in joints of masonry to be exposed or painted shall be raked out as necessary, filled with mortar, and tooled to match existing joints. Immediately after grout work is completed, scum and stains which have percolated through the masonry work shall be removed using a high pressure stream of water and a stiff bristled brush. Masonry surfaces shall not be cleaned, other than removing excess surface mortar, until mortar in joints has hardened. Masonry surfaces shall be left clean, free of mortar daubs, dirt, stain, and discoloration, including scum from cleaning operations, and with tight mortar joints throughout. Metal tools and metal brushes shall not be used for cleaning.

3.21.1 DRY-BRUSHING

a. Exposed concrete masonry unit
b. Exposed concrete brick surfaces
c. shall be dry-brushed at the end of each day's work and after any required pointing, using stiff-fiber bristled brushes.

3.21.2 CLAY OR SHALE BRICK SURFACES. Clean exposed clay or shale brick masonry surfaces as necessary to obtain surfaces free of stain, dirt, mortar and grout daubs, efflorescence, and discoloration or scum from cleaning operations. After cleaning, examine the sample panel of similar material for discoloration or stain as a result of cleaning. If the sample panel is discolored or stained, change the method of cleaning to ensure that the masonry surfaces in the structure will not be adversely affected. The exposed masonry surfaces shall be water-soaked and then cleaned with a solution proportioned 30 mL 1/2 cup trisodium phosphate and 30 mL 1/2 cup laundry detergent to 1 L one gallon of water or cleaned with a proprietary masonry cleaning agent specifically recommended for the color and texture by the clay products manufacturer. The solution shall be applied with stiff fiber brushes, followed immediately by thorough rinsing with clean water. Proprietary cleaning agents shall be used in conformance with the cleaning product manufacturer's printed recommendations. Efflorescence shall be removed in conformance with the brick manufacturer's recommendations.

3.22 BEARING PLATES. Set bearing plates for beams, joists, joist girders and similar structural members to the proper line and elevation with damp-pack bedding mortar, except where non-shrink grout is indicated. Bedding mortar and non-shrink grout shall be as specified in Section [CAST-IN-PLACE CONCRETE].

3.23 PROTECTION. Protect facing materials against staining. Cover top of walls with nonstaining waterproof covering or membrane when work is not in progress. Covering of the top of the unfinished walls shall continue until the wall is waterproofed with a complete roof or parapet system. Covering shall extend a minimum of 600 mm 2 feet down on each side of the wall and shall be held securely in place. Before starting or resuming, top surface of masonry in place shall be cleaned of loose mortar and foreign material.

3.24 WASTE MANAGEMENT. Manage waste according to the Waste Management Plan and as follows. Minimize water used to wash mixing equipment. Use trigger operated spray nozzles for water hoses.

3.24.1 SEPARATE AND RECYCLE WASTE. Place materials defined as hazardous or toxic waste in designated containers. Fold up metal banding, flatten, and place in designated area for recycling. Collect wood packing shims and pallets and place in designated area. Use leftover mixed mortar as [retaining wall footing ballast] [cavity fill at grade] [underground utility pipe kickers] [_____] where lower strength mortar meets the requirements for bulk fill. Separate masonry waste and place in designated area for use as structural fill. Separate selected masonry waste and excess for landscape uses, either whole or crushed as ground cover.

3.24.2 TAKE-BACK PROGRAM. Collect information from manufacturer for take-back program options. Set aside [masonry units, full and partial] [scrap] [packaging] [_____] to be returned to manufacturer for recycling into new product. When such a service is not available, local recyclers shall be sought after to reclaim the materials. Submit documentation that includes contact information, summary of procedures, and the limitations and conditions applicable to the project. Indicate manufacturer's commitment to reclaim materials for recycling and/or reuse.

3.25 TEST REPORTS

3.25.1 FIELD TESTING OF MORTAR. Take at least three specimens of mortar each day. Spread a layer of mortar 13 to 16 mm 1/2 to 5/8 inch thick on the masonry units and allowed to stand for one minute. Prepare and test the specimens for compressive strength in accordance with ASTM C780. Submit test results.

3.25.2 FIELD TESTING OF GROUT. Field sampling and testing of grout shall be in accordance with the applicable provisions of ASTM C1019. A minimum of three specimens of grout per day shall be sampled and tested. Each specimen shall have a minimum ultimate compressive strength of 13.8 MPa 2000 psi at 28 days. Submit test results.

3.25.3 EFFLORESCENCE TEST. Test brick, which will be exposed to weathering, for efflorescence. Schedule tests far enough in advance of starting masonry work to permit retesting if necessary. Sampling and testing shall conform to the applicable provisions of ASTM C67. Units meeting the definition of "effloresced" will be subject to rejection. Submit test results.

3.25.4 PRISM TESTS. Perform at least one prism test sample for each 465 square meters 5,000 square feet of wall but not less than three such samples shall be made for any building. Three prisms will be used in each sample. Prisms shall be tested in accordance with ACI 530/530.1. Seven-day tests may be used provided the relationship between the 7- and 28-day strengths of the masonry is established by the tests of the materials used. Compressive strength shall not be less than [_____] MPa psi at 28 days. If the compressive strength of any prism falls below the specified value by more than 3.5 MPa 500 psi, steps shall be taken to assure that the load-carrying capacity of the structure is not jeopardized. If the likelihood of low-strength masonry is confirmed and computations indicate that the load-carrying capacity may have been significantly reduced, tests of cores drilled, or prisms sawed, from the area in question may be required. In such case, three specimens shall be taken for each prism test more than 3.5 MPa 500 psi below the specified value. Masonry in the area in question shall be considered structurally adequate if the average compressive strength of three specimens is equal to at least 85 percent of the specified value, and if the compressive strength of no single specimen is less than 75 percent of the specified value. Additional testing of specimens extracted from locations represented by erratic core or prism strength test results will be permitted. Submit test results.

3.25.5 SINGLE-WYTHE MASONRY WALL WATER PENETRATION TEST. Prior to start of field construction of the single-wythe masonry wall, perform masonry wall water penetration test on mock-up wall assemblies consisting of the identical design, materials, mix, and construction methods as the actual wall construction and in accordance with ASTM E514/E514M. Prepare a minimum of three specimens and cure for minimum 28 days prior to testing. Construct panels by the same methods, processes, and applications to be used on the project's construction site. The spray test duration shall be 6 hours for each specimen. No water shall be visible on back of test panels during the test and any areas of dampness on the backside of the test panels shall not exceed 25 percent of the wall area. Dampness is defined as any area of surface darkening or discoloration due to moisture penetration or accumulation below the observed surface. Construct additional test panels for each failed test performed until three test panels pass the test. Factors that can affect test performance include materials, mixing, and quality of application and workmanship. Materials, mixing, and methods adjustments may be necessary in order to provide construction that passes the water penetration test. Document and record the test specimen construction materials and application and provide written test report in accordance with ASTM E514/E514M, supplemented by a detailed discussion of the specifics of test panel construction, application methods and processes used, quality of construction, and any variances or deviations that may have occurred between test panels during test panel construction. For failed test panels, identify in the supplemental report any variances, deficiencies or flaws that contributed to test panel failure and itemize the precautions to be taken in field construction of the masonry wall to prevent similar deficiencies and assure the wall construction replicates test panel conditions that pass the water penetration test. Submit the complete, certified test report, including supplemental report, to the Owner prior to start of single-wythe masonry wall construction. Significant changes to materials, proportions, or construction techniques from those used in the passing water penetration test are grounds for performing new tests, at the discretion of the Owner.

www.ingramcontent.com/pod-product-compliance
Lightning Source LLC
Chambersburg PA
CBHW081902170526

45167CB00007B/3113